Monday February 2, 2039

I'm second again, behind Katie Martin I got the silver award – I already have too many of these second place awards. I need to work harder to run harder. My brother Garrett takes advantage of every silver medal to tear at my ego piece by piece. When I arrive home I see my mom cooking Shelby, Garrett, and I our dinner. Spaghetti again – the second place meal of my life. I was always promised that if I ever got first I would get a steak dinner. Only once had this occasion happened. My sister Shelby has to be hand fed because she cannot control her body movements, she embarrasses me more than anyone can imagine. It is freezing outside here in Hazelton, Pennsylvania with 14 inches worth of snow, once again. My name is Rosie Conley, I am in the eighth grade and I love to run! Today at school my teachers are having us write about this asteroid that is supposed to come really close to the earth in the next few days. My science teacher had us write formulas for the asteroid, my history teacher had us discuss the important part of history we will soon be witnessing, and to top it of my math teacher made us calculate how far the asteroid would be from the earth. This is the first page of my diary, and I assure it will not be the last, welcome to my diary.

Saturday February 6, 2039

The asteroid is supposed to come by today everyone is so excited. Garrett got his telescope out today, and my mom brought out her old opera binoculars. Shelby could not do anything, but sit around and wait for it to be nighttime. I am not excited for this event as much as other people are. I don't think that the asteroid will even be visible, just like every other meteor shower that never appears. Our family had a vote to see what we should do to celebrate and watch the asteroid – Garrett voted for a late night picnic, Mom voted to go to the lighthouse and watch from there at New York's border, I vote nothing I simply stared off into space just as Shelby.

Later tonight, 9:00 pm - I am even a little bit excited to see this, so I brought out a lawn chair to the picnic outside in the backyard. CNN news stations are going crazy with astronomers because the asteroid is supposed to pass by any minute now, it's calculated time is 9:03 pm. The time ticks slower than ever as the anticipation is killing me. We all gasp as the clock gets to 9:03. There it is!!! A huge floating ball of rock…coming *towards* us, that isn't supposed to happen!! Terror strikes the earth at that very moment, we hear screams from local neighbors and praying begins, all at that very moment. This is when we hear a big BOOM; the asteroid had broken the sound barrier, and was rushing towards the earth. I looked inside our house, and saw the TV screen had gone scratchy. We ran inside to our house, and listened to radio stations, they didn't have much more knowledge than we did just that an asteroid had struck the Earth. The radio then announced that the asteroid had hit somewhere in the Pacific Ocean, there was a massive tidal wave that was going to destroy everything from California to Kansas. There would be massive weather conditions, and the outcomes would be ***catastrophic.*** NASA had miscalculated the asteroid, and because of that mistake half of the Earth would be dead in a matter of minutes or hours. Mom told us that *we* had nothing to worry about because we are nowhere near Kansas. She told us that we would still have school on Monday, and that these reports seem sketchy so there is nothing to worry about. We saw that she was hiding something though, something that we had never seen before in mom – fear.

Very late that night, we went to bed or should I say laid in bed. I am now thinking about what would happen next, I had seen the science channel before, and that was what scares me the most. We are not prepared for a doomsday disaster like some people that actually believed in that kind of stuff were. Tomorrow morning I would help mom look for solutions, help her more than I ever had before because this was a very stressful time for everyone especially her.

Sunday February 7, 2039

I do not know of one person who did not go to church today. The church was filled with people praying for loved ones, innocents, and praying they would not become one who died.

After a 4 hour long church mass, we heard rumors of NASA preparing a massive living ship to go into outer space that could carry us all to safety because we would all die soon if we stayed here. The ship was supposedly going to take off in New York near the border of PA and NY. This was not too far from our house, so we could probably be one of the 6,000 families that could board the ship. People think there will be another one taking off in Michigan as well, but we should stick to this one. The ship was supposed to take off at 6:00 am on Tuesday. We all had to remember that these were rumors because the electricity was gone, and there was no more TV or email by computer anymore. We had to listen to these rumors for life's sake, and a little bit of hope in our daily lives.

Monday February 8, 2039

No school today. Julie, my best friend, came over, and I told her all about the ship departing from the border. As if she hadn't heard it already. She said that she would tell her family, I told her not to tell anyone else but her family because the more people who knew, the less chance we had of being on that ship.

This afternoon mom packed our bags and told us we would be leaving tonight so that we could get into line for the morning. The border was only 1 hour away and I had no idea what the woman was talking about getting there a whole night early. I had to go along with it just as Garrett and Shelby did too. We packed plenty of food from the local market, tons and tons and more tons of food, we packed a lot of our clothes, and so many heating sources, flashlights, blankets, pillows, personal hygiene things, hundreds of batteries and pills for every kind of disease you could get. We had a total of 16 suitcases by the time we were done packing. We loaded the car and got on our way.

Once we got the station we saw that we would check-in and then unload our bags in a certain room, 213 on the left wing was our room. The ship looked like a massive hotel, I was surprised how big it actually was and how NASA actually pulled it off! There were already 2,784 families there and I searched for Julie's family name – Kovenger, no sign of it. I hope she arrived soon or she wouldn't get on! It wasn't like I could text her because there was little to no phone service. The people at the front of the ship gave us each one device that could text each other messages as long as we had the code numbers for the device, without using cell service, I thought that was awesome! The ship would serve buffets every night, so much for the food we brought. The first 3,000 families got everything free, so we were so glad. I felt bad for the other 3,000; I hope that Julie's family is one of the next 3,000 because the spaces are filling up fast. The lady at the front desk made mom fill out a form about what she majored and minored in college. Even though mom was a doctor, she lied and said that she was a teacher. She had heard rumors that if you were a teacher than you would be given special advantages, perhaps a bigger and more convenient room.

Our room was equipped with 2 bedrooms, a bathroom, a living space with 2 chairs and a couch, a kitchen that ran on steam, and a room filled with seats connected to the floor and the wall. Immediately we started to unpack the food because it might be hard to unpack with the turbulence we could come upon. It took us all at least 1 hour to pack the kitchen with the food. There was plenty of cabinet space so the food fit well. Next, we started to unpack our clothes – I would be sleeping in a room with Shelby, ugh. Shelby won't take up much room because she does not care about how she looks, therefore more drawers for me. I sorted my clothes into drawers by the following categories: tops, bottoms, undergarments, pajamas, and other. Then we decided to put a flashlight, a heater, two extra blankets, and two boxes of tissues in each room. I decided to quickly head downstairs and check the list for Julie's family. The number of families by now had skyrocketed to 4,206. My finger shook as I slowly moved down the list towards the K's. No sign of Julie. I could see at least 13 more families in line waiting to board the ship. When I passed the B's I thought I saw a familiar name, Befard, I wonder if it is someone I know? I ran back upstairs to our room, and told mom that there was no sign of Julie. She said that there was still plenty of space left, but I really hope that Julie hurries up.

At 10:31 pm we went to our rooms to go to bed. It was a long day for all of us; once I knew everyone was asleep I quietly left the room for a total of 3 minutes to check the list. All 6,000 families had boarded. I checked the list one last time I thought my legs were going to lock on me and I would make a big scene. I slowly moved down the K's: Ka, Ke, and Ki...Kovenger!!! She had made it, room 839 on the left wing. I went back to the room and saw that no one had noticed I left, I have I feeling I will sleep well tonight.

Tuesday February 9, 2039

At 4:00 this morning the speaker woke everyone up telling us to eat our breakfast and to be ready to take off at 5:45 am. Our family speed walked up to the restaurant that was only a few rooms away, what a convenience. There were restaurants filling the entire ship. I ate scrambled eggs and two biscuits. After a nice hour and a half breakfast, we went to the ship's arcade.

Garrett seemed to be having the time of his life. It seems that life here won't be too bad.

5:30 – we went up to our room and waited for the speaker to tell us what to do next. Soon enough the speaker came on and told us to sit in the

take off/ turbulence/ landing room that was equipped in our room. We were to fasten every one of the six seat belts on the seat. The speaker then announced that the trip up to space would be about 200 seconds, but the ride would be rough the whole way. We were also told there could be a loud sound if we hit the sound barrier fast enough. When the speaker started the countdown from twenty, things got really quiet among us all. Shelby had to be strapped in with her wheelchair and I felt like it was not very secure. 10…9…8…7…6…5…4 the engines started spinning rapidly and I grasped the seat with my knuckles surely as white as they could be. I suddenly felt sick as the ship turned upright on the count of 3. The speaker started the three second countdown 3…2…1…blast off! Just as you had heard in the movies the ole blast off saying in every space movie. I don't quite remember anything else that happened in the next 5 minutes, all I know is that we passed out because of how fast the ship was moving. We now know why astronauts have training before this traumatic experience.

The rest of that night the bathroom was filled with us throwing up because of the experience. We were not floating because the ship had gravity machines on board. It is a good thing that we packed plenty of saltine crackers and ginger ale because we all needed it tonight.

Wednesday February 10, 2039

After a long night last night throwing up and going to the bathroom we made it through the trip to space. Early in the morning I woke up and decided to walk up to room 829. Elevators don't work in space so there were a few flights of stairs I had to hike. When I got to Julie's room and knocked on the door it opened rapidly before me. There stood Julie looking terrible, but probably no better than I looked. We hugged and cheered and laughed and then hugged some more. Julie told me that a space school was supposed to open next Monday, I usually would not want that, but it makes me feel more civilized and makes me feel like none of this had ever happened.

Thursday February 11, 2039

I fell down a flight of steps today while we were heading downstairs to lunch. I was taken to the ship nurse. At the office it was like an infirmary with IV's, much like a hospital back on Earth. Going there made me feel secure about living on a spaceship because I knew that we had plenty of medical supplies. I found out that I had sprained my ankle and that I have to wear a brace and wrap for 2 weeks, and that I should stay off it for the majority of the time. I got ice cream from mom because she felt sorry for me and I got a teddy bear from the gift shop.

Friday February 12, 2039

It was a boring day today – all I did was sit around and learn how to knit a sweater with the tons of yarn mom packed for times like these. So far I have knitted a low quality blue sweater and an even lower quality red.

Sunday February 14, 2039

This weekend was even more boring; the highlights of my day were talking to Julie about rumors on the ship.

Monday February 15, 2039

Today school started and there were about 12 more eighth graders on the ship so the school decided to combine grade levels 6th, 7th, and 8th together for classes which gave us a total of 32 students. I felt bad for the 6th graders because they would never really get a sixth grade education. We were told that we would be learning 7th accelerated and 8th on level classes. The classes available to us were Math, Science, Language Arts, History, Gym, and Choir or Art. We could choose whether to take art or choir. I chose choir because I have always wanted to learn how to read music for a hobby. My brother chose art so we agreed to get two of everything so that we could both participate in everything. They would serve Pb &j, pizza, or macaroni everyday for lunch at school. Today was really easy because all we did was get our schedules and introduce ourselves to our classmates. I met three really nice girls; Cassie, Hayley, and Jordan. Cassie is in eighth, Jordan is in sixth grade, and Hayley is in 7th grade. I have the same schedule as Julie and Cassie; I have two or three classes with Hayley and Jordan. My schedule was pretty simple – Language arts for two periods, Gym, Science, Math, History, and then Choir.

Friday February 19, 2039

School was okay this week. I have nothing much to say about school besides the fact that I am learning things I learned last year. In choir we learned how to sight-read music, it is really fun but kind of hard too. We also got music for the Green concert in March. The pieces we got are called, "A New Day" and "In Time of Silver Rein".

I was assigned the part of a Soprano I, the highest part of the songs. Mom told us not to eat too many snacks in between meals because of our health, but I knew that meant she didn't want us eating up all of the food.

Monday February 22, 2039

Everybody on the ship forgot about Valentines Day a week or two ago, so the school decided to have Valentines Day today! Everyone brought in handwritten valentines with candy attached and for lunch the cafeteria gave out a pink or red cupcake for celebration. I ended up with a pretty big load…I got a total of 18 valentines. Julie gave me a whole present including 2 boxes of chocolate, 14 hard candies, a pack of lollipops, and to top it off a cute teddy bear. Cassie gave me a whole pack of peppermints, Jordan gave me 3 packs of sweethearts, and Hayley gave me 4 pink and red bows to wear to school. One or two kids gave a goodie bag full of sweets and little toys. Mom gave me a beautiful heart necklace once I got home. All the others consisted of simple "Happy Valentines Day" and a jolly rancher attached. I gave out chocolates to my outer friends, and gave goodie bags full of stuff to my close friends, but for Julie I got a valentine sweater and a net that you throw a ball at and it rebounds back at you. I also included some tennis balls to play with. She said that once I heal that I could come over and play on the net anytime. So, today was not a totally horrible Monday as they used to be. Im glad this ship has a series of shops to go to, otherwise nobody would have had anything for Valentine's Day to give out.

Wednesday February 25, 2039

Tuesday whizzed by because the homework that Mrs. Kay gave out was a really fun assignment. She gave us a scavenger hunt around the ship; it was a little harder for me so she gave me a revised list that consisted of things on my floor and the main floor.

Thursday February 26, 2039

I got my brace off today! I ran over to Julie's room and we played with the net I gave her, my ankle still feels sore. I think it is because I haven't moved it in a while, but I am going to keep an eye on it.

Friday February 27, 2039

After school today I slept over at Julie's room for a little celebratory fun night, Cassie came over too. We watched movies, played board games, and pranced around in our pajamas around the hotel until curfew. By 11:00 pm we were all out dead in sleep.

Sunday February 29, 2039

We went to church early this morning for mass. We haven't been able to walk very far with me, and the church is on the 11th floor and there was no way I could go up the stairs a few days ago, so we had to skip church. It felt good to pray with everyone today.

We passed by Mercury,and we all gathered around a glass room to see it. It was a huge tourist attraction so we stopped traveling for today at about noon. I realized then that the reason we were sent up here was to find a habitable place to live because by now everyone on earth had died for sure. There was no way anybody could survive the asteroid.

Saturday March 6, 2039

This week went by so slowly, I could not write in my journal at all because of all the homework I have had. Sorry for the long break in time.

The choir concert is on Thursday and I am so excited for it! Progress Reports came out yesterday, and I have all A's with all of them higher than 97. I got a free ice cream coupon because of my great grades from school. Nearly all of the 6th graders failed math because they were not taught integers and such. Julie told me today that the ship was having upper and lower class divisions tomorrow because we were running out of supplies like pillows and blankets. She said the requirements were not very hard so we should be able to get into higher class.

Sunday March 7, 2039

Julie was right…today we had upper and lower class divisions. Both of our families were put in the upper class division. We had to go to a huge meeting room on the main floor of the space ship today to talk about the changes and revisions while on the ship.

We were able to have Brunch and Dinner because the food brought had seemed to have a shortage. They also said that the lower division families were moving in the basement so that the ship could have some more space for waste. The higher-class families would be moved to floors 1,2,3,4, and 5. We should be fine because we were already on the 2nd floor, but Julie would have to unpack and move rooms.

Tuesday March 9, 2039

Today at school Mrs. Kay pulled us out of all of our other classes to practice for our concert on Thursday. She says that the concert is really important because everyone is invited, even lowers. We needed to be really good to bring joy to everyone on the ship. She also gave us one more song to learn because the longer the concert, the better. There are 3 choir classes, and we all have 3 different songs; Women's

(Varsity), Men (Varsity) and Mixed (minors). I was put in the Women's Varsity because there was not a lot of Soprano Is. Since each choir has 3 songs, this means that we will be sining 9 songs in all. We will also be serving snacks, I think the lowers and maybe even the highers will love the celebration. Everyone is noticing some tension between the lowers and highers because the lowers are jealous. There are signs and people are going on strike from their jobs. "THE HIGHER CLASS DOES NOT HAVE TO WORK – WE SHOULDN'T EITHER".

Julie's new room is #273 it is pretty close to our room so it made us both really happy when we saw how close it really was. I made her a green sweater for St. Patrick's Day a while back when I hurt my ankle, and I forgot when St. Patrick's Day actually was so I gave it to her today after school.

Thursday March 11, 2039

Today was the choir concert and it was a big hit. Lowers came and made it a real party. We had cookies and all of the choir kids wore green dresses and green tuxedos. I was surprised they had enough material to make all of the outfits; considering they were ugly it must have taken them 2 minutes to do each one of them.

We are starving, and I can't imagine how the lowers feel with only one meal a day. I heard they were about to send people out into space to try and find edible things to eat. We are still doing "okay" because we packed a lot of food.

We can't let anyone know how big our actual stash is because they might break in or try and come to our room to eat it all.

Friday March 12, 2039

Today at school they gave us a really hard test. People are saying that they gave it out because they wanted to see if any kids were smart enough to go out into space to find food. Surprisingly my brother did really well on the test. He passed the first 3 rounds at ease and then they gave him documents to memorize and we found out that he has a photographic memory. Of course NASA is going to want him in space with them so he can tell them all kinds of cool facts. So on Monday he would be sent out to space with some of the smartest people there was on earth. We are all devastated because there is a high chance that we will never see him again. He was given botanist books to memorize by Monday so that they could know the signs of whether things were poisonous or not.

Monday March 15, 2039

This weekend was really rough on all of us. Garrett departed at 6:00 am this morning, the time is now 4:00 pm. I could not bring myself to write about Garrett leaving any other time sooner. We were all just huddled up together crying with him in his spacesuit and Mom and I in our pajamas. We packed him his favorite snacks in his backpack, and told him we loved him with our souls. We assured him that we would see him again, and that if you love something to let it go and it will come back to you. And as quick as that was, I didn't have a brother anymore and mom didn't have a son anymore.

I didn't go to school today because I was so devastated.

Friday March 19, 2039

This week was really hard for all of us. We decided to go to the ice cream store today to cheer ourselves up. Mom and I discussed the fact that we would see Garrett again because he was with terrific men and women that would watch over him. Once he got back home we would never let him go again.

Wednesday March 22, 2039

We are all starving and my family has been pushed back to the lowers.

People have started killing others to eat. Lowers are getting one meal/snack everyday. The food we have in our room has been taken out for the higher-class people.

We are living in the basement of the ship, during the nighttime I am scared, everyone is scared. The school has closed down because somebody burned it down on Monday. I have no secrecy anymore, and somebody wrote the following in my journal that I just had to erase, "*Be Afraid*"

What kind of a person would write that...living on this ship is terrible. People are starting to look at Shelby like a snack because she won't see it coming. Mom and I are doing our best to protect her, but sometimes I see a little chicken around her too.

Tuesday March 28, 2039

We all knew this was coming, but I was so shocked when I actually saw it. When I was coming from the bathroom/corner I saw a dog pile of men and women on something. They seemed to be around our unofficial territory so I chased them off. When they left a few minutes later...I saw Shelby's wheelchair next do this pile of bones and leftover tissue.

Shelby had been killed and eaten...

Wednesday March 29, 2039

Mom and I are trying to look at the bright side to Shelby's passing...She won't be a hassle to us anymore? That just made us cry harder.

Friday March 31, 2039

Today our ship got hit with a meteor that took off the left wing. Half of our supplies, half of our people, and half of our remaining oxygen. There is nothing but a sheet covering the gaping hole.

We are all going to die sooner or later.

Tuesday April 4, 2039

There is a sickness going around these past few days. People are saying that it is the flu. I have a strong immune system and so does my mom, but we cannot last forever. Each lower got one suitcase and all the rest got taken away along

with our room. In my suitcase I packed 4 sets of clothes, making sure I could run and be free to move in them. I also packed some snacks that I have been keeping for emergencies and my journal. I packed baby wipes, tissues, two stuffed animals I had gotten on special occasions on the ship that reminded me of special memories and a flashlight. I also packed a brush and all the ponytail holders i had brought on the ship, which ended up being not as many as I thought.
Finally, I packed a tiny heater and a tiny fan.

Mom brought a huge suitcase; it was extra, extra large. She included 4 sets of clothes for her, All the medicine we brought, flashlights, a little bit of food, water bottles, and two blankets and a pillow. Since every lower family got an air mattress mom and I both slept on the mattress and used the two blankets to keep us warm. I never really used a pillow back at home anyways so I let mom have the pillow. She also brought yarn, tape, and 4 sheets; this was our system of curtains.

Wednesday April 5, 2039

Julie's mom got the flu...even though we are friends and all, we did not give her our medicine because if we got sick we would need all of the medicine we could get.

Times are tough these days.

Friday April 7, 2039

There was nothing we could to save Julie's mother...she was already dead by the time we woke up.

Julie is devastated, we told her to bring her mattress to our space and to live with us from now on. She does not have a father, much like me, this was how we became best friends because we both had to deal with our fathers dying. My mom would take her mothers place, but we all knew there would be nothing to make her as bright as she was a few days ago.

Sunday April 9, 2039

No church today. No food today. No life today. Nothing.

Monday April 10, 2039

You don't realize how precious life is until you almost lose it. A guy had somehow smuggled a knife on board. Since I run long distance it was easy for me to run for *my* life. Jordan, however…there is no amount of words that can explain her long and painful death today. It was during nighttime that it happened, I awoke to a shrieking scream, and I figured it was Jordan by morning. Mom told us that we had to run and hide, she brought our suitcases and we made a run for it. Thankfully Julie ran with me and my mom was in great shape so we all made it to a safe place in time. We soon heard a shot out of a shotgun and we peeked around the corner. The knifeman was dead. Next to him we saw a mound of dead people, I started looking at faces to see if any of them were familiar. All I could do at that very moment was pray that I didn't know anybody, but I did…Jordan was near the bottom of the mound. She was one of the first killed in that terrible massacre of bones and blood that day.

I am writing this on Tuesday because I could not force myself to write yesterday, it was terrible, truly inhumane.

Thursday April 14, 2039

Today we had our weekly memorial for all the people who died in the past week. This memorial was really long because many people had died in the knifeman's massacre.

They did it by alphabetical order – Jordan's last name was Ferdin, and of course Julie's mom's last name was Kovenger so we listened for their names to be called and prayed for them.

It was a terrible process first we would scrap the muscles off of a person, strip them straight to the bone. We would store the muscles in bags to be eaten later on. Next, we would stack the bones on a mound, once everyone's bones were on the mound we would say a few prayers and people would take them someplace and open up the waste hatch. The bones would float off into space along with their spirits. We all cried when Jordan and Julie's mother was stripped because of the memories we had with them.

Monday April 18, 2039

They have a list posted of the people who will be terminated for food in the next week; so far none of us are on any of the lists.

Befard was on the list, which made me wonder more about who he was. I decided to find out in the next week before he would be terminated on Friday.

Tuesday April 19, 2039

I got Julie in on the investigation today. She said that she had heard that name before as well.

This afternoon or some time around then, we have lost track of time, Julie and I asked around about Befard. Some people were saying that he was a boy and looked about 14. We asked if anyone had seen him in the past few days, everyone said no. We looked everywhere on the 3 places us lowers were aloud to go on; the basement, the trash room, and the exit. He must be on the run because he was nowhere to be found. I cannot imagine the highers would notice him because the highers couldn't have been living a much better life than we were.

Wednesday April 20, 2039

We decided that we would go ahead and go up to the highers territory today. We cleaned ourselves up with the wet wipes I packed earlier, and put on the least dirty of the clothes I packed. I picked a purple T-shirt with a pair of purple sweats. I did my hair with the brush and the ponytails I packed earlier in my suitcase they came in handy. We ended up not looking too shabby...Julie even had earrings in.

When we got to the stairs the service man told us that lowers weren't aloud on the stairs. I said the following in a tone that Julie had clearly never heard before,

"DO WE LOOK LIKE LOWERS TO YOU, THOSE HALF DRESSED PEASANTS!! WE CAME DOWN HERE TO OBSERVE THE LOWERS FOR A SCHOOL PROJECT!! AND YOU BETTER NOT ASK US FOR OUR HIGHER IDS BECAUSE THAT IS A HUGE INSULT TO ALL OF THE HIGHERS! I CAN GET YOU FIRED FROM THIS JOB IN AN INSTANT!"

That certainty shut him up because he immediately let us in the doorway to the stairs. He turned around in a quick manner and actioned for us to pass.

We looked all around for a 14-year-old boy. We finally spotted one about 30 minutes in. We asked him for his name and he wasn't Befard, but he did tell us that he used to go to his school and he said that he had an old photo of him. He gave us the photo and told us that he wouldn't need it back. We gladly took the photo and continued on with our search. The main floor was the biggest floor so it would take the longest. We realized then that he probably wouldn't be somewhere so popular. We then decided to go up to the abandoned top 5 floors on the right wing. We would check every room.

Once we finally got to the top floor we started checking rooms. I would check the take off room, the bathroom, and the kitchen. Julie would check the bedrooms, living rooms, and closets. After about 4 hours worth of searching we finally found him on the third to top floor. Julie's face was as red as a cherry when she realized who he was. It was her ex-boyfriend, Craig. It was the most awkward moment when he said to Julie,

"do I know you?"

Craig is not the brightest 9th grader, and he certainly doesn't have a very high IQ at all. I decided to use the bathroom for the first time in nearly one month. The room clearly had a spider problem, and the bathroom floor had a hole in it that was growing mold, but it is still a bathroom. I would do anything to go back to life, as it was where the bathroom was an everyday luxury. When I came out, Julie and Craig were making out. I quickly ran downstairs back to the basement.

I am writing this at night on my mattress with my mom, she doesn't seem to mind the flashlight in her face, so why not. I know I shouldn't have just left Julie there, but I heard her make her way through our curtain soon after I had fallen to "sleep". Nobody can sleep on this death ship anymore; it is too dangerous to sleep.

Friday April 22, 2039

Finally some good news!! Garrett came home today, and he got his own air mattress because of how important everyone thought he was. He got a bag packed with goodies filled to the top because of his heroic attitude! The bag included; snacks, hygiene stuff, a tiny tub that you could wash your feet in, but we would use it as a bath tub, and a pass to go up to the main floor to get food and water for our family! He also brought home a tiny cage, or as we called it at first a dresser. It was really a doghouse for the rescued puppy he had found when he was on the ship. The puppy had somehow gotten on to the ship before take off and he told the others that he would take care of the little one. The others agreed and one very nice woman gave him a cage for the little puppy. The puppy's name is Sherman, and we already love him. He is a yorkie, and will always and forever fit into that cage. Sherman would have to eat human food, but he would be in heaven every night! We told Garrett that tomorrow he would get Sherman a food bowl (paper bowl), a water bowl (paper bowl), a washcloth, some mini diapers, and a tiny waste bin with a lid for us to keep Shermans trash inside of. This was the first family pet our family ever had, and we are so excited to care for Sherman.

Craig was murdered today.

Saturday April 23, 2039

Julie and I play with Sherman nearly 24/7. Some of our close neighbors come in our curtain to play with Sherman too!

Friday April 29, 2039

We are getting well feed again, and some of the other lowers (who are left) are noticing and glaring at us in a terrifying way. Sherman is good, otherwise life is great. Sherman prances around our space, which we enlarged, and wears his tiny diapers that collect his waste.

Friday May 6, 2039

Today is the 4-month anniversary of us in the grasp of the devil.

Monday May 9, 2039

Today we passed by Venus, and we all had to use oxygen masks that fell from the ceiling when we passed it, but it was a breathtaking view.

Nearly all of the lowers have died, so they are sending some more people down here on Friday, we are survivors.

Monday May 16, 2039

The past week was really boring, but it was really emotional for the new lowers. Our new neighbors are nice, so we told them that they could have some food if we ever have any leftover, that is considered super nice in these days; that would have been considered rude in the old Earth days.

On the other hand, Mom and I have a hypothesis about Julie. You know how a few weeks ago Julie made out with Craig, we think she is pregnant. She is getting round and she is throwing up. She runs to a corner and then comes back all light headed and dizzy. I think I will tell her I know in a few days after the news isn't so shocking to her.

When Craig was murdered she was so emotional, and now I know why…she is scared because there will be no father to take care of her newborn. I am not a mind reader, but I know that Julie doesn't want her baby having to experience a life without a father like she did.

I wonder if she is considering abortion. I know that she never believed in abortion but these are desperate times. I wonder if doctors are even offering abortion anymore, I wonder if doctors even exist anymore. I have a feeling if you went to the front office because of a minor foot injury now a day that all of the supplies would be used up because of the flu outbreak we had about a month or two ago.

Wednesday May 18, 2039

Mom had a long talk with Julie about her pregnancy, just like Julie's original mother would have talked about. I didn't eavesdrop because I didn't want to hear a second of it, so I left our curtained off space.

Thursday May 19, 2039

I just found out that my friend Cassie died a few days ago on the 14th. I am so sad because we had a special connection.

Friday May 20, 2039

I had a tiny bath today, and it felt so good. Julie packed soap a while ago, and it came in handy today. I feel so clean. It feels awesome to be clean for once. Mom did my hair and made me look gorgeous. I put in Julie's earrings (secretly). I snuck up to the higher class, and I had a blast because I looked old enough to get in the nightclub. It was the best party I have ever had in my life. There was so many pretty lights and endless music. Apparently a lot of pop stars got on board because New York City is not to far from the boarding place back on Earth. Justin Beiber was on the ship, Usher, Rihanna, Adele, and finally Blake Shelton. It was awesome!

When I came back to the space my mom was in on the whole idea and told me that I deserved it for one night of my life. She was happy for me, she was happy that I was doing something fun for once since we got moved to the lowers.

Thursday May 26, 2039

It was mom's birthday today. Garrett bought mom a sweatshirt that said

"I am a happy person."

It really made us laugh until we were in tears. Julie said that she felt the baby kick today; I don't think it is that big of a deal.

Hayley peeked around our curtain today and said hi. She was the longest of us all to stay in the higher class. She was in the past movement of lowers this past week.

Friday May 27, 2039

Hayley opened up to us today and said that she was living alone as of right now because her family was infected with the flu a month or two ago. She also said that the higher-class life isn't much better than this.

We told her to move her mattress to our space. Instead she told us that they were giving out cots instead of mattresses. We told her that that was still okay, and she said that would be a blessing. In that case we expanded our space, again.

In her suitcase she had packed many blankets, *some* food, and 6 sets of clothes. She told us that we could all share clothes since we were about the same size anyways.

Saturday May 28, 2039

Today when I went to the bathroom corner I saw a bunch of bones in cages, next to the bones I saw slaves carving things out of them. You can only imagine whose bones they are. Actually now that I am thinking about it, at Jordan's funeral a week or two ago we never actually saw the bones being thrown out of the ship. The reality of all of this is that they never threw them out and that at any moment other people's bones could kill my friends, family, or myself. People who are on the lists now don't even get killed in their sleep. They get killed wide-awake and afraid.

Sunday May 29, 2039

Today our tiny church group got bigger by a family or two. We praised the lord and sang together.

Tuesday May 31, 2039

Mom is teaching us Science and Math because she majored in science and minored in math because back on earth she was a scientist. She lied on the papers on the ship and told them that she was a teacher because she did not want to abandon our family. We learned how to tell if things are poisonous.

Friday June 3, 2039

Some guy took a bone spear today and somehow got up to the higher class and killed anyone in his way while getting up to the captain's cockpit. Once he got to the cockpit he snuck up on the captain and killed him. It was a slaughter; far more than a slaughter to us it was a life or death moment.

We were now lost in space due to the fact everyone that could possibly drive the ship to our safety is dead. Nobody thought it could get any worse, but it just did.

Saturday June 4, 2039

Everyone is in a panic, our snack bag is still plenty filled, and Garrett still has his pass, but it is still hard for us.

We told Garrett to find some non-expiring foods and to bring back all the food he could carry.

The other day Garrett found a pack of water bottles not being used and he brought them back to us and we washed our clothes. It feels so good to have a clean set on right now. The people on the ship typically give each lower family a bowl of water every 2 days some crackers and beans every day, that's it.

Sunday June 6, 2039

Today my enlarged family gave hope to others. We spread cheer. We told them all about funny stories and encouraged them to tell us theirs. We laughed and shared for hours. Some next-door neighbors came after a little while and we had a merry time.

Wednesday June 8, 2039

We have heard rumors about how we passed something called a nebula today. It threw us off track, like we had any sort of direction. We were supposedly headed towards Neptune at a faster pace. We didn't feel any change in movement, but it was because these last few days had been rough with turbulence.

Thursday June 9, 2039

The flu is back; apparently it was never really gone in the first place.

Julie has the flu. She is sick and looks so weak and pale. We gave her Tylenol and so much other medicine, but none of them seem to be working. We need help, but nobody is around to help. The population has gone from 6,000 happy families to about 350 deprived families.

We pray everyday, but our main goal is to let Julie die in peace or in her sleep. We want her to be happy when she dies, which we all would like to be sooner rather than a long painful death. When she passes, I will be by her side.

Friday June 10, 2039

Julie is in pain…everyone is too shell-shocked to do anything but watch the horror.

Saturday June 11, 2039

Julie passed last night in her sleep. As sad as it is the baby died as well. There will be no baby. We are glad it had to happen like this, but we are very sad. Julie was my best friend and has left a forever-visible footprint in my soul. I knew I had to do something that I would have never done back on earth. I am going to hide Julie's bones after the memorial. She will not be a weapon in her afterlife, she will be happy.

Sunday June 12, 2039

It was Julie's funeral today. I cried for hours after the funeral. After looking at the bones that were in the case I took them out and put them in some families abandoned sheet. I wrapped them up and snuck up to the garbage shoot.

Nobody even bothered stopping me, they were all an emotional wreck up in the higher class, so I went on my way.

I gave Julie a proper speech and said great memories. I told her goodbye and I let her go out of the shoot. I told myself that it was the right thing to do, but I still felt like things would be better on earth than here. I wonder what is happening on Earth.

I wonder if anyone survived, I wonder if they are making progressions, I wonder if life is good there.

Monday June 13, 2039

I probably haven't mentioned Sherman in a while now. He is still perfectly fine and healthy. We made up his birthday to be July 4. About a month ago Garrett found a squeaky toy in the lobby of the ship. We happily are saving it for Sherman's birthday!!

I am now proud to announce he is potty-trained! He goes to the bathroom in a little box in the corner. Nobody cares that Sherman is here, I used to have to hide him, but to be honest everyone is just too busy protecting food and loved ones at the moment rather than to care about a mutt puppy.

Saturday June 19, 2039

This week went by really slowly. I finally found my messenger thing that I got at the beginning of our trip. I had lost it a long time ago! I messaged my mom while in the bathroom corner just to see if it worked. Sure enough I got a message back in about 2 minutes tops saying

"We're sorry your message(s) could not be delivered. Please try again later."

The messenger didn't work anymore, of course not, just like any other thing that NASA has done for us in the past 6 months.

Sunday June 20, 2039

We feel that we are doing the ship a favor just by being alive. So today we decided to move up to the 1st floor rooms. There would be no one there to stop us and nobody would care. We encouraged all the other lowers to come up with us too. About ¾ lowers agreed with us and came up to living again. Apparently our ship ran into a satellite or something about 3 months ago?? Nobody knows for certain, but this is what Hayley tells us, she also says that a lot of people had died in the wreck because a lot of carbon dioxide was produced. Our oxygen quality has gone down quite a bit I noticed right away. I felt like the closer I got to an actual room; the more I could see, the more I could hear, the more I could feel. When my hand touched the doorknob to our new room my emotions rattled inside of me and created more butterflies than anyone could count.

I am writing in a room on a bed with Hayley. I have decided to let Hayley read and write in my journal in her free time. She loved the idea and so there might be some point of view changes in the next few chapters, but the more detail the merrier, and I am happy to share the journal with her.

Hayley is now like the sister I never had.

Monday June 21, 2039

Hey, this is Hayley. I don't know if anyone will ever be reading this because there might not be anyone left by the time this is found, but whatever. It is fun anyways, something to do.

This is me again, Rosie Conley.

Today we didn't really do anything except for sleep in a bed. I haven't slept in a bed for months now! Here are the sleeping arrangements; Hayley with me in one room, mom and Garrett in one room, Sherman has the kitchen all to himself.

I wore Julie's signature jeans today. I tried to think of it as honoring her, but I knew it was wrong at the same time. I wish I could just see her again, laugh with her again.

Friday June 25, 2039

Today I had a fight with Rosie; she wouldn't let me wear her shirt because she wanted to wear it. I mean how rude is that! She didn't even say sorry afterwards. I think she wants me to apologize, but I have nothing to say to her.

How dare she think it is acceptable to write in <u>MY</u> journal after she won't even apologize for being a downright jerk! She clearly did not seem to recall the fact that the person who owned the article of clothing has first dibs on it. We had that set in the rules clearly, but I am not apologizing. She needs to first.

Saturday June 26, 2039

Hayley and I are friends again, but it took some explaining to do. I had to have her brain remember that whoever's article of clothing it is shall have first dibs.

I think she gave me a little payback today though. She wouldn't let me wear her sweatpants this morning, and I think that little taste of payback gave her the right to be my friend again. Neither of us ended up apologizing, but we kind of just became friends again. There was no peace treaty; we just decided that we are better as friends than enemies. Hayley says that she probably won't right in my journal very often, because she is going to find one in a room. I am still confused at this very moment about what she meant, "in a room". She couldn't have meant our new room?

Sunday June 27, 2039

We prayed a little. We ate a little. We lay around a lot.

Monday June 28, 2039

The school had closed down, but mom certainly hasn't. She gave us tons of information on plants today. She keeps going all nuts that we will have to know this someday.

I found out that Hayley had an idea that we could ransack old apartments and see if they had any food or anything for us in them. Sadly, we thought it was a great idea. We are going to cover 5 rooms each tomorrow. I am rooms 234,235,236,237,238, and 239.

Tuesday June 29, 2039

It was fun to ransack old rooms because you come across some of the weirdest things. Today I found some really valuable stuff though. Foodwise I found; old cereal, old fruit, and 12 cans of beans.

Cloth Wise I found; 4 pairs of sweatpants, 2 sweatshirts, and 3 shirts. I also found a box of matches stashed away in an old drawer. There must have been a little one living there because in every room there were toys lying around everywhere. I decided to take 3 home for Sherman; I would give them to him on his birthday, which is coming up.

Hayley found old cheese its and 9 cans of corn. She also found some jewelry to accessorize us. There is no more buffet as of tomorrow so

Garrett was out all day today, just going down and carrying up all he could and going back down. He repeated this step as many times as he could before it was time for the buffet to close down forever. By the end we had tons of food, so much food!! It is awesome, mom did ten rooms and she found food galore in one of the rooms. She found a board game called twister. We played that a lot throughout the night tonight.

Wednesday June 30, 2039

Mom told us that from now on we would fast on and off every day. She said that we would have brunch and dinner. She also told us that on Mondays, Wednesdays, and Fridays that we would be eating only lunch.

We weren't thrilled about the idea, but it was our obligation to accept.

Thursday July 1, 2039

Today we ate well, for dinner we had beans and rice. For our drink we had some tap water in a bowl. We decided that on Friday we would search for more cups and such.

We search on the days that we fast to take our minds off of the fact he we are hungry. It really did help on Wednesday.

Friday July 2, 2039

Today I searched 5 rooms and found; plates, silverware, and cups for the kitchen. I found 24 cans of peas all in one room. It was a score!

For my personal liking, I found a new journal. Once this one is used up I will have a new one. I am actually almost done with this diary so I am really glad that this stroke of luck happened. I heard that mom was supposed to go see if any of our old stuff was still in our old room. She said she would go on Sunday.

Sunday July 4, 2039

Independence day, Sherman's Birthday, and Old room day!!! To celebrate mom said that we could eat today. We had a feast at dinnertime; we ate baked chicken, mashed potatoes, beans, corn, and some bread. I don't know how mom pulled it off but she sure did it well. To celebrate Sherman's birthday we wrapped his 4 presents in a blanket. We also gave him the blanket to lie on. He seemed to have a wondrous time with his squeaky giraffe, squeaky carrot, tennis ball, and his plastic bone that would last him years to finish. He was so happy!! Our room had some food in it, some water bottles, and flashlights. We have had no electricity in a long time by now. We lost it about 2 months ago and it hasn't came on ever since. We have had to live by flashlight and sometimes sunlight from different suns. Today was the best day I have had on the ship ever since March.

Monday July 5, 2039

Somebody tried to come in to our room today. I think they were trying to ransack our apartment, but luckily mom always keeps the doors locked. Today we took a break from stealing because we were very tired. Our oxygen is really bad quality right now, so it is harder to breathe. We need to get off of this ship and go to somewhere safe that we can live on.

The only problem is that we don't know where we are going, or how to drive a spaceship!

We told Garrett to go up in the cockpit on Wednesday while we are searching rooms.

Wednesday July 7, 2039

Garrett came back this afternoon and said that he knows some basic maneuvers like how to turn and somewhat land. He also said that there are so many buttons that he could push, but he doesn't want to experiment with anything quite yet.

Thursday July 8, 2039

There was a black hole today that our ship almost ran into. It was so scary; it was like this big rotating black mass. We ran into the turbulence room and hugged together. Garrett ran up to the cockpit and turned the ship towards the left side and saved the day. Everyone thinks he is a hero, who he is, but now they want him to be the captain of the ship. Mom asked Garrett if that was what he wanted, and he said that he would love to be the new captain of the ship. The captain gets a huge suite right next to the cockpit, so we would be moving again.

The suite has a huge lobby with a stage on it, and a piano on the stage. It is a 3-story suite and is so cool. My room has a built in set of bars to do gymnastics on. I was never really into gymnastic but it is still something to do. It has a vending machine in my room, only it is preloaded with $597 to spend. Hayley's room is right across from my room, so we share a little living space in between. Garrett's room has a built in trampoline, you can tell the old captain had a big family and was filthy rich. Hayley's room had a beautiful modern taste to it. It has a nice white couch and many white fluffy and so soft bean bags and it has a canopy bed. I love the new suite. Sherman's room came with huge bags of dog food and a soft and cuddly beanbag. Everything seems to be going great, and I am so thankful that this all happened to us. We are living in a great family right now, we might not be a legal family, but we are a happy family. Hayley and I are great friends, and I hope that it stays that way. She is like the sister I never had.

Thursday July 8, 2039

Mom went swimming in our personal swimming pool today, she told us that the pool felt nice, and that we should go swim. That is what we did, we went to the pool in our room, and we swam in our underwear. Nobody cared anymore, nobody was even there. Hayley felt a little awkward at first, but she got used to it. It was just like old times back on earth. The classic chicken fights, the screaming, the fake water stunts, all of it. Here I am sitting on the comfy couch writing about how good our lives are right now, meanwhile there was an enormous murder today for food. 23 families were killed in the massacre this afternoon. That brings that brings the population down to 233 families alive. My mom told us some breaking news today as well. She told us with a serious face on, that we need to get pregnant. She said the population is low and it is our job to keep the human race alive.

I shall not explain what happened tonight with Garrett and I, but I do not like it one bit! This will be the night that changes my life forever, because I am about to become a mother, in 5 "sweet" months.

From now on I am only going to write important things that happen on the ship every week or so, at least until I have my baby.

Saturday July 17, 2039

Today we passed by Neptune; this was not supposed to happen. Neptune is the farthest planet away from the earth besides the Dwarf Planet Pluto. We believe that the thing that hit earth was actually Neptune. It had been following us after it made the impact on earth. If this really did happen we knew that the earth would be dead by now, no signs of life for sure.

Neptune looked terrible with a bluish shade and it looked like it had been drenched in a blue goop.

Saturday July 24, 2039

Today my mom told me that she knows that I will be okay once I have my baby. She also told me that our family has a history of earlier babies, and that my baby will probably be born at 3 months, I thought that was terrible because if that does happen neither my baby or me will be safe. One of us could even die during birth. My baby will need extra care, and most likely my baby will need extra supplements. I will accept my baby no matter what. My mom found a doctoring book a few weeks ago while ransacking, and she gave the book to Garrett to memorize, he really comes in handy now a days. Population on the Death Ship: 187.

Sunday July 25, 2039

I know I am not supposed to write everyday from now on, but I just had to write about this. Today 12 more surviving families came to our church group. We have set a date for caroling, Sunday August 8. Summer caroling seems to have the right idea for staying bright on the idea that we are lost in space. We are going to sing the following songs to bring joy; Alleluia and Halleluiah. Garrett is also doing really well on controlling the ship. He has learned how to get rid of some of the turbulence we have been experiencing. Things are getting better, but people are still dying for food.

Sunday August 8, 2039

Today we went caroling, and people joined us as we went along. We ended up nearly having the all the survivors sing with us. It was a joyous day because in the end we all ended up singing in the lobby like a giant chorus. Tears of joy started streaming down mom's face because at that moment we thought that everything was going to work out. We would work together to get through this. We are a big family; brother: sister, mother: child, daddy: daughter.

Saturday August 21, 2039

Today my baby kicked.

Garrett fainted from lack of oxygen. There really isn't that much oxygen on the ship anymore. If we can't find more oxygen soon we won't make it through to see my baby.

Reality speaking: What is the point of all this? What were we really sent up here for? Is this all really worth being tortured through life and afterlife? How can we make this all better?

Saturday August 28, 2039

Garrett told us that he thought he saw another planet from his telescope, but he very well maybe lying. He said that it was hidden behind Venus before, and we couldn't see it from the angle we had been at from planet earth. He said that even if it was a planet he didn't know how to land this ship on a planet. I walked up to the cockpit and took a look at it for myself. It wasn't very visible, but I did see that it was green and blue, much like earth. It also had what looked to be white puffy stuff. I thought that it was earth at first, but then I realized that Earth probably wasn't recognizable anymore. I can only imagine the humanity of earth, the big dent in the earth's surface, the massive weather conditions, the massive deaths and casualties. I can't believe that life as we knew it had been completely terminated, life as we knew it was gone.

Wednesday September 2, 2039

The baby is getting close to due date. So today my mom and I discussed possible baby names. If the baby is a girl I will name her Leia, if the baby is a boy we will name him Scott.

My mom told me that since there wouldn't be any medication while I have my baby that there is a lower chance of survival of my baby or me.

Friday September 11, 2039

This is Laura, Rosie's mother. She had her baby today. It is a baby girl and we named her Leia. Her baby is a very special baby. Leia gives something to the ship, something that no scientist could fix nor find. Leia gives the ship hope. She is very healthy and carries hopes of a great future.

Rosie will return to her writing on September 20. She is very weak and ill. She needs to rest.

Sunday September 20, 2039

I am still very weak as my mom mentioned before in the other entry before this one. I am not even hungry, but I am still better than before. While Hayley was making her daily scavenge hunt in the rooms she found a little carrot. We gave it to Leia and she played with it all day.

Tuesday September 22, 2039

I am better now, and I am so glad. I got to play with Leia today, and she is so cute. I am so glad

I have my very own baby. Just like I had imagined as a young girl. All the baby dolls and baby bottle toys had paid off because I can already tell that I am going to be a great mother. Leia has blonde hair with beautiful blue eyes and a smile that nothing could ever replace.

Wednesday September 23, 2039

Since Leia has no clothes I asked if Hayley and mom could go look for some baby clothes. They said they would love to, and I am so happy that I have a dependable family.

Thursday September 24, 2039

Hayley and mom had no luck finding any baby clothes. I guess we'll just have to cut adult clothes to make baby clothes. I am pretty sure that there weren't any babies aloud on the ship before hand. There certainly wasn't any baby food on the ship either. We would have to steal fresh fruit every morning and chop the fruit up in the blender for every meal. Our process seemed to have some flaws in it, because we would always end up leaving the lid off the blender!! Just like we did today, haha it was funny though. Although it was a pretty big mess.

Friday September 25, 2039

I went scavenger hunting this afternoon, but I only did 3 rooms instead of five. It is hard work lugging a bag full of things around the ship.

Today we decided to use our swimming pool, and I swam in some old clothes I found in another room. I also found a huge score in one of the rooms I scavenged, a battery operated sewing machine! I started sewing some clothes for Leia as soon as I got back into our suite. Leia now has two sets of clothes, and one pair of footies. One set of clothes is pink, and the other is green, and her footies are baby blue my favorite color.

Saturday September 26, 2039

We went ahead and scavenged again, and this time found nothing, except for an old toy train. I am saving up some things for Leia for Christmas. Santa *will* make his way up here in an unidentified position.

Sunday September 27, 2039

Our church group met today in our suite. One family brought chips and quesò to eat. I met a new friend that came to our room, her name is

Skylar. She hoists her hair up into cute Dorothy brown braids. She wore a blue and black dress, and some nice flats. We could already tell that she was one of the most fortunate of us all. She always has a grin on her face, and I don't know what to think of that. She has green eyes, the color of emerald. They shine in the far beyond sun. Her mother is very weak, so she came to our church group today. She told us today that her mother would probably die; we let her stay for dinner. She has very pleasant manners, and knows all of her please and thank you's. Although she seems like a tomboy, always having a tennis ball with her. She is really nice as well. We sent her home about 2 hours after dinner.

Tuesday September 29, 2039

She knew it was coming, but Sky's mom died today, and she is very sad. The ship doesn't even have any more funerals; they just strip the bones of muscle. Sky took all of her belongings and moved in the suite with us. We have always had a second story ever since we have moved to the new suite, but nobody ever bothered to go up there. Since there are not enough rooms down here for Sky, I am going to move upstairs with her. That way she doesn't feel so awkward sleeping up there alone. Tonight she just slept on the couch, but tomorrow we are going to explore the upstairs.

Apparently the population is down to 79 families. All that is left of planet earth is 79 families.

Wednesday September 30, 2039

Today we all explored upstairs, it has 3 bedrooms, a workout room, and a playroom. The playroom was already equipped with a play kitchen and a rocking horse. Leia doesn't know about it yet, but I am going to surprise her with it for Christmas. My room must have had a girl about my age living in it beforehand. It had blue walls with green polka dots. My room had a bed that was white and a canopy bed, and the dresser was a 4 shelf. I decided to organize my clothes by undergarments, shirts, bottoms, and other. It even had a mirror; those were hard to come across these days! It also had a white desk. In the drawers of the white desk it had a ton of paper, colored pencils, crayons, tape, scissors, and stickers. In another drawer of the desk it had ponytail holders, a brush, some headbands, and bobby pins. I decided to put my diary lock on that drawer. I also put my diary in the drawer before I closed it. I was the only one in the family that knew the combo to that lock. I made sure to have a lock on my diary when I got it at the age of 10.

In between Sky and I's rooms we have a living space that has a comfy couch and 2 comfy chairs. I am so happy right now, I feel like this whole ship was a nightmare, and now it is a dream. There are also extra oxygen tanks in the living room that we brought downstairs incase of an emergency where we start running out of all oxygen. There are like 17 oxygen tanks in all that are supposed to last for 2 months.

Wednesday October 1, 2039

I woke up and brushed my knotted hair for the first time in ages. I will probably keep this brush hidden because it is a true treasure these days. I drew a fake window. Outside the window in my drawing was just like what my view from my room at my house was. I made it have a white picket fence with the street sign saying Hazelton right outside in plain view. I went to Sky's room and together we explored the drawers of the family room that is in between our rooms. In the drawers we found a D.S. with 6 games and its charger was solar powered and we were so close to so many suns that it would never run out of power. The games were Super Mario Bros, Pet Palace, a Plane game, a car racing game, Picture Editor, and Carnival Games. We decided that on Monday, Wednesday, and Friday that I could play with it, and the other days would be Sky's days, except on Sundays, nobody would get it then. In another drawer we found a ring of keys. We don't know what they are to yet, but we will probably find out soon. The key ring probably had about 34 keys in total. Sky and I then decided to figure out what the keys went to.

In our room we have extra oxygen in tanks because we are so "special" to the ship. There is a really low supply of oxygen left on the entire ship, if Garrett doesn't find someplace for us to find oxygen and live then we all might die up here. Then the whole human race will be extinct, it is up to us to keep it alive.

Friday October 3, 2039

Leia gave her first words today she joyously blurted out, "ma ma, and ma ma".

It made me sob to think I am a mother now my whole life has changed.

Garrett says that the thing he saw earlier was actually just a glob on his machine and that there is no other planet around that we can live on.

Saturday October 11, 2039

Sky and I found a mysterious door that is located on the top of the ceiling. We don't think that it is the attic because we already discovered that, but it seems creepy. So we told everybody that we would go up there tomorrow all together, Leia, mom, and all.

Sunday October 12, 2039

After we had our weekly service with everyone left on the ship we told people to follow Sky and I up to the door. One of the only men left on the ship grabbed the string to pull the staircase down, but it was locked. And he couldn't break it down. After about 30 minutes of trying to break down the door I remembered the key ring and handed the ring up to the tallest man to unlock the door. He tried ring after ring but nothing happened. We told them that we would get it all sorted out by the next service, and to come back next time. I have no idea how mom even thinks that we are going to get that door open. It is steel and it is sealed shut.

Monday October 13, 2039

I just thought I might explain to you how the key ring looks. It is more like an ancient stone than a ring. I had a mythically drawing embroidered in the middle. Then almost right to the edge it had a huge circular hole that went all the way around the outside of the rock. This is what held the rings.

Sky and I went to where the door's lock seemed to be and tried to peek through but we couldn't see anything it was pitch black inside. Then I accidentally tripped and the rock ring hit the side of the door to where the drawing made an imprint on the door. Then all of the sudden the door opened. We ran downstairs to get mom. She was so surprised when she heard the news she ran to the drawer and grabbed three flashlights for all of us to carry. We rushed back up the stairs into the room and slowly turned on our flashlights. Once we went in further we got more scared because we realized that anything could be in here a psycho killer, bodies, or even worse. Then my mom all the sudden screamed a heart wrenching scream and both Sky and I turned our heads and flashlights to her and found out that she had just ran into the light string that was hanging from the ceiling. We turned on the big light and saw all kinds of majestic chests. Each chest had a key lock on the front of it and we opened up one chest and it led to a secret passageway. We climbed down the ladder that was in the chest and then we took a sharp left and found another chest, and we did the exact same thing that we did before about four more times. Until we came to a door, and when we opened that door there was a bright purple circular thing spinning. Mom told me to gather up the ship and to meet her back down here. So I did and everyone was surprised and squeezed into that tiny room. Nobody knew what the portal was to. We had to decide who would be the tester of the portal. Nobody raised his or her hand, and I did the most stupid thing I have ever done in my entire life, I raised my hand.

People cheered, and my mom screamed, "NO, DON'T DO IT ROSIE YOU COULD DIE OUT THERE!!"

I couldn't make out any words through my mouth I felt like I was about to pee my pants. I felt a tear drip down my face. I hugged my mom and she wouldn't let me go. Next, I held Leia possibly for the last time in my lifetime. Then I stood up and walked towards the brightest light I have ever seen and felt my hand being grabbed and being pulled back. Then I leaped toward the light and that leap is what changed my life forever.

www.ingramcontent.com/pod-product-compliance
Lightning Source LLC
Chambersburg PA
CBHW030902180526
45163CB00004B/1669